The AI Handbook for Seniors and Their Caregivers

ALICE CUTHBERTSON

Published by **Aliboo Publishing**
www.aliboo.com

For permissions, collaborations, or inquiries, please get in touch with aliboo@aliboo.com

DEDICATION

To all the seniors who carry wisdom, stories, and love that the world needs to hear
and to the caregivers whose patience and compassion give dignity to every day.

And to my mom, whose work as a nurse gave me the gift of walking the halls of a nursing home as a child.

Because of you, I learned early the value of listening, connecting, and cherishing the lives of our elders.

CONTENTS

PREFACE

When I first began writing this book, I wanted to create something that would feel like a trusted companion for seniors and caregivers. I wanted every chapter to be clear, supportive, and free of the confusing jargon that too often makes technology feel out of reach. But here's the truth: I didn't do it alone. I used artificial intelligence itself as a partner in the writing process as well as memories, loved ones and my intent.

AI helped me edit my drafts, and organize ideas. When I struggled to find the right words, AI suggested options. AI helped me simplify. It was like having an always-available editor and brainstorming partner one that never grew tired of reviewing a paragraph for the tenth time. Importantly, every word in this book still carries *my voice and my vision.* AI did not write for me it worked with me. I asked the questions, provided the stories, and set the tone. AI acted as a supportive assistant, offering suggestions that I could accept, reject, or reshape.

I chose to share this openly because I believe it demonstrates exactly what this book is about: **AI is not here to replace us; it's here to support us.** Just as AI helped me bring clarity and compassion to these pages, it can also help seniors remember appointments, stay in touch with loved ones, or rediscover joy through music, stories, and photos.

This book is not only about AI it is also a your intended collaboration with it. And my hope is that knowing this will give you courage to try it yourself. If AI can help me write a book, imagine the ways it might help you in your daily life.

INTRODUCTION

Most people have never experienced what I did as a five-year-old girl, running freely and gleefully through the hallways of a nursing home. The place was packed with residents of different ages, states of health, and stages of life. I remember the giant corridors, slick and slippery, carrying the sharp scent of bleach mixed with something funky I couldn't name back then. It was a smell that belonged only to that place.

I would run, I would skip, and as I passed the rooms, I'd see all kinds of scenes: some people awake, some asleep, some moaning, some calling out. A few would notice me darting by and call out, "Hey, come back, little girl!" At first, I would slow down, peering in cautiously. Almost always, there were bright eyes waiting for me, a gentle smile, a soft beckoning: "Come talk with me."

I was there because my mother was a nurse, and without daycare, she brought me along to work. Those residents became my companions. Some visits were joyful, some carried deep sadness, and all were temporary, because, for some, leaving meant passing to another realm. Even as a child, I felt the weight of that truth. I often thought about how disconnected they must feel, some without visitors at all, others with occasional ones, each separated in their rooms except for the rare social event or visit from a volunteer.

The other day, those memories came rushing back. The wisdom I gleaned from those departed souls, the kindness in their eyes, and their hunger for connection suddenly tied itself to an idea: *What if they had a way to learn about*

AI? What if we could offer them not fear, but empowerment, something tangible that could make their lives richer, easier, and more connected?

And so, this book was born. It is the second in a series I have committed to share with you, a handbook not just for parents of children, but for caregivers of seniors and for seniors themselves who are still capable, curious, and ready to thrive in this new digital age. You don't need to be a "tech person" to read this book. You don't need to know business jargon, complicated settings, or the inner workings of computers. This handbook is written to feel like a friendly conversation across the table. Each chapter is filled with plain-language explanations, real-world examples, and simple steps you can try at your own pace.

My hope is that you will set aside any fear, lean into curiosity, and discover that AI is not here to replace you nor leave you behind, it is here to support you. Whether you are a senior yourself or a caregiver helping someone you love, this book will be your companion as you explore technology made friendly.

1 CHAPTER: MEET YOUR NEW COMPANION

Most of us grew up in a world where "technology" meant the radio, the television, or maybe the family's first rotary phone. Now we hear a new word everywhere: **AI**, short for *artificial intelligence*. For many seniors, this word feels intimidating, as if it belongs only to scientists or science fiction movies. But here's the truth: AI is already quietly working alongside you every day, and it isn't nearly as mysterious as it sounds.

Think of AI as a *helpful companion*. It's not a person, and it's not here to take your place. Instead, it's a collection of smart tools designed to make life easier. When your phone corrects a misspelled word, when closed captions appear on your television, or when your GPS gives you directions, that's AI lending a hand.

AI in Plain Language

AI is simply a way of teaching computers to "learn" from information. Just as you might learn from reading the newspaper or practicing a skill, computers learn by being given examples. The difference is that they can look at millions of examples very quickly and then make suggestions or predictions.

For example:

1. Netflix recommends a new movie based on what you've watched before.
2. Your phone suggests the next word as you type a text message.
3. A voice assistant like Siri or Alexa answering, "What's the weather today?"

None of these tools is "thinking" in a human way. They are following patterns to give you helpful answers.

Why Seniors Don't Need to Fear AI

The idea of AI can feel overwhelming because it's invisible; you can't see the gears turning. But you don't need to see the wiring in your refrigerator to trust it will keep your food cold, and you don't need to see the inner workings of AI to benefit from it.

What matters is this: AI can **help you stay connected, keep you safe, and add joy to**

your life. And you are in control; you decide when and how to use it.

Caregiver's Corner

If you're reading this as a caregiver, think of yourself as a guide, not a technician. Your role isn't to explain every detail, but to *translate technology into comfort.* Encourage small wins:

Start with simple success, like helping someone use voice-to-text instead of typing.

Offer patience. Many seniors carry the fear of "breaking something." Reassure them it's safe to explore.

Celebrate progress, no matter how small. A successful video call or a well-timed reminder builds confidence.

A Gentle First Step

Let's try something together:

1. Pick up your phone or tablet.
2. Find the microphone icon in your text

messages.

3. Tap it and say, “Hello, this is my first AI message.”
4. Watch your spoken words appear as text.

That’s AI in action, simple, supportive, and right at your fingertips.

In this chapter, we’ve met AI not as a threat but as a companion. In the next chapter, we’ll explore how it already surrounds us in everyday life, quietly lending a hand in ways you might not even realize.

NOTES

2 CHAPTER: AI IN EVERYDAY LIFE

Artificial intelligence might sound like something hidden away in research labs, but the truth is, you've probably used it dozens of times today without even realizing it. AI is no longer a futuristic idea; it's woven into the small, everyday conveniences that make life run smoother.

Where You Already Use AI (Without Knowing It)

Television: When closed captions automatically appear on a show or movie, AI is interpreting speech in real time.

Cell Phones: Predictive text suggests the next word as you type. Your phone's camera sharpens the photo automatically; that's AI.

Email: Spam filters keep unwanted messages out of your inbox.

Banking: Fraud detection alerts when something unusual happens on your account.

GPS: Maps give real-time traffic updates and suggest faster routes.

Smart Speakers: "Alexa, play my favorite song." AI finds it instantly.

You've been living with AI for years. It's not something to fear; it's simply the "invisible helper" built into the tools you already trust.

Everyday Benefits for Seniors

1. Reminders: AI can ping you when it's time to take your medication or pay a bill.
2. Accessibility: Tools like voice-to-text let you speak instead of type, which is easier if your hands are tired or your eyesight is changing.
3. Entertainment: AI can suggest music you loved decades ago or find old TV shows you haven't seen in years.
4. Connection: Video calls become clearer and easier to manage with AI smoothing the background noise.

Caregiver's Corner

Caregivers play an important role in making these tools approachable:

Translate the invisible: Instead of saying "this is AI," try saying, "Your phone has a helper that remembers things for you."

Simplify setup: Program the first reminders or voice commands together, so seniors feel success right away.

Spot opportunities: If a senior enjoys music, show them how to ask a smart speaker to play a favorite song. If they worry about forgetting things, introduce simple reminder apps.

Try This: Familiar Faces

If you've ever used Facebook or a photo app, you might have seen how it recognizes faces in pictures. That's AI, too   it sorts and groups people so you can easily find photos of your grandkids or old friends.

Open your photo gallery.

Search for a family member's name.

Watch as your phone collects all the photos of them into one place.

This is AI making organization effortless.

Reassurance: You're Still in Control

AI doesn't take over your life. It listens for your input, offers suggestions, and then waits for your decision. You can always ignore its suggestions, turn it off, or choose something else.

In this chapter, you've seen that AI isn't far away   it's right here in the everyday. In the next chapter, we'll look at one of the most important ways AI can help: staying connected with family and friends.

NOTES

3 CHAPTER: STAYING CONNECTED

One of the most powerful gifts AI offers is helping us stay connected with the people we love. For seniors, connection is more than convenience; it's a lifeline. Whether it's talking to grandchildren across the country or checking in with a neighbor next door, AI makes those conversations easier, clearer, and often more fun.

Simple Ways AI Helps You Connect

Video Calls: Apps like Zoom, FaceTime, or WhatsApp use AI to adjust sound, brighten images, and keep your face centered.

Voice-to-Text Messages: If typing feels uncomfortable, simply speak, and the words appear as text.

Translation Tools: If your family speaks more than one language, AI can help bridge the gap with instant translations.

Noise Reduction: Background sounds like barking dogs or a TV in the other room can be softened automatically so your voice comes through clearly.

Everyday Scenarios

A grandmother uses voice-to-text to send a

quick "Love you!" message without struggling with small keys.

A grandfather joins a family birthday celebration through a video call, where AI keeps his face bright even in a dimly lit room.

Caregivers set up group messages so daily check-ins are easy and stress-free.

Caregiver's Corner

If you are supporting a senior:

Set up shortcuts: Put the call button for family members right on the home screen.

Do trial runs: Practice a call or message before it's "important," so seniors feel confident.

Encourage independence: Show them once, then step back, let them press the button, and experience success.

Reassure often: Remind them it's impossible to "break" the app by pressing the wrong button.

Try This: Send a Voice Message

1. Open your text messaging app.
2. Look for the microphone icon.
3. Press and hold it while speaking: "Hello, this is my first voice message!"
4. Release the button and hit send.

   You just used AI to turn your voice into a message.

Reassurance: Connection Brings Confidence

The more seniors experience these small wins, the more their confidence grows. AI becomes less of a mystery and more of a bridge, a tool for laughter, love, and the warmth of human connection.

In this chapter, you discovered how AI helps keep relationships strong. In the next chapter, we'll explore how AI can be a wellness partner, helping with health, safety, and everyday peace of mind.

NOTES

4 CHAPTER: YOUR HEALTH & WELLNESS PARTNER

Good health is the foundation for living well, and AI can quietly become a supportive partner in maintaining it. For seniors, AI isn't about replacing doctors or nurses; it's about providing **gentle reminders, helpful tools, and extra confidence** in daily routines. For caregivers, it offers peace of mind, knowing there are extra safeguards in place.

How AI Supports Health Every Day

Medication Reminders: Apps can chime when it's time to take a pill and even track whether it was marked as taken.

Appointment Alerts: AI calendars send nudges for doctor visits, therapy sessions, or family events.

Exercise Support: Fitness trackers count steps, encourage movement, and even suggest gentle exercises.

Nutrition Guidance: Smart apps recommend recipes based on dietary needs (low-salt, diabetic-friendly, heart-healthy).

Cognitive Wellness: Brain games, puzzles, and memory apps powered by AI can keep the mind active and engaged.

Everyday Examples

1. A senior receives a friendly reminder on their watch: "Time for your afternoon walk."
2. A phone app flashes a notification: "Take one pill with water now," preventing missed doses.
3. A grandson sets up a digital grocery list with his grandmother, and the app suggests healthier substitutions automatically.

Caregiver's Corner

For those guiding seniors:

Start Small: Begin with one helpful feature, like a medication reminder, before introducing more.

Pair It with Routine: Connect reminders to daily habits (e.g., pill alerts at breakfast time).

Check In Together: Review data from fitness

trackers or apps during visits. It turns health tracking into a conversation, not a chore.

Use AI as a Backup, not a Replacement: Stress that technology supports but never replaces human care.

Try This: Simple Reminder

1. Open the calendar app on your phone.
2. Add an event: "Drink water" at 11 a.m.
3. Set it to repeat daily.
4. Watch as AI reminds you right on schedule.

That little nudge might feel small, but over time, these reminders build healthier habits.

Reassurance: AI is a Partner, not a doctor

It's important to remember that AI is a **helper, not a healer**. It won't diagnose illness or replace professional advice. Instead, it's a friendly partner that helps you stay on track, stay safe, and feel more in control of your daily health.

In this chapter, you've seen how AI supports health and wellness with small, simple steps. In the next chapter, we'll tackle one of the biggest concerns for seniors and caregivers alike: **staying safe in a digital world.**

NOTES

5 CHAPTER: SAFETY WITHOUT FEAR

One of the biggest fears seniors express about technology is: "What if I get scammed?" or "What if I press the wrong button and break something?" These worries are real, but here's the truth: you can stay safe with a few simple habits. Just like locking your doors at night or looking both ways before crossing the street, digital safety becomes second nature once you know what to look for.

The Most Common Digital Risks

Scams & Phishing: Fake emails, calls, or texts pretending to be from your bank, the government, or even a family member.

Misinformation: Articles or videos designed to look real but spread false claims.

Identity Theft: When criminals try to use your personal details for fraud.

Unwanted Charges: Accidentally signing up for subscriptions or services you don't need.

Simple Rules for Staying Safe

1. Pause Before Clicking: If a message or link feels "off," don't open it. Ask a trusted family member or caregiver first.

2. Never Share Personal Info by Text or Email: Banks and Social Security will never ask for your password or Social Security number online.

3. Check the Source: Look at the email address, website link, or caller ID closely. A scam often looks almost   but not quite   real.

4. Use Strong Passwords: Simple phrases work best, like PurpleTulip1960! instead of "1234."

5. Turn On Two-Factor Authentication (2FA): This adds an extra code sent to your phone for logins, giving double protection.

Caregiver's Corner

For those helping seniors stay safe:

Create a Safety Checklist: Post a simple "Do's and Don'ts" card near the computer or phone.

Practice Together: Show examples of scam emails vs. real ones. Confidence comes from recognition.

Set Boundaries: Encourage seniors to always "check with you first" before making large online payments.

Celebrate Safe Choices: When a senior avoids a suspicious email, recognize that as a success!

Try This: Spot the Scam

Ask yourself three quick questions before clicking on anything:

Do I know the person or company sending this?

Does the message sound urgent or threatening? (Scammers love pressure tactics.)

Is there a spelling error or odd link?

If something doesn't feel right, delete it or ask for help.

Reassurance: You're in Charge

The good news is, you have more control than you think. Scammers rely on fear and confusion. But when you pause, ask questions, and use simple tools, you take away their power. AI itself is even helping fight scams by filtering suspicious calls and emails, so in many ways, technology is on your side.

In this chapter, you learned how to recognize risks and protect yourself online. In the next chapter, we'll switch gears to something lighter: the fun side of AI how it can bring joy, play, and creativity into your daily life.

NOTES

6 CHAPTER: AI FOR JOY

Technology isn't only about health and safety; it can also bring delight. AI has a playful side that can spark creativity, stir up memories, and even bring a good laugh. Joy matters at every age, and AI can be one more way to invite it into daily life.

How AI Can Add Joy

Music on Demand: Ask a smart speaker to play songs from your teenage years or a favorite artist for instant nostalgia.

Storytelling & Games: AI can tell you a bedtime story, play trivia games, or create jokes.

Art & Creativity: Programs like AI painting apps can turn your photos into drawings or create pictures based on your words.

Memory Lane: AI photo apps can enhance old pictures, bring faded colors back to life, or make slideshows of family photos.

Everyday Examples

A grandmother asks, "Alexa, play Elvis Presley," and the living room fills with the King's voice.

A caregiver sets up an AI trivia game after dinner, sparking laughter and memories around the table.

A senior uploads a faded wedding photo into an app, and it's restored with clearer colors and details.

Caregiver's Corner

For those supporting seniors:

Lean Into Interests: If your loved one enjoys gardening, try an AI app that suggests plant care tips. If they love poetry, let AI write a new poem together.

Make It Social: Use AI-powered games or music as shared activities, not solo ones.

Keep It Light: Avoid overwhelming features, focus on one fun tool at a time.

Encourage Experimentation: Remind them it's

okay to play around. There are no wrong answers in creativity.

Try This: AI Trivia Night

1. Ask a voice assistant: "Tell me a trivia question."
2. Take turns answering with your family or friends.
3. Let the AI keep the fun going with more questions.
   It's simple, silly, and surprisingly engaging with just your voice.

Reassurance: Fun Builds Familiarity
Every laugh, song, or shared moment with AI also builds comfort and familiarity. The more positive experiences seniors have, the less

intimidating technology feels. Fun becomes the doorway to confidence.

In this chapter, you discovered AI's lighter side: music, stories, photos, and games that bring joy. In the next chapter, we'll explore how AI can be used to strengthen family bonds and bridge generations.

NOTES

7 CHAPTER: STRENGTHENING FAMILY BONDS

Family is at the heart of life's joy. For seniors, staying involved with children and grandchildren keeps spirits high and combats feelings of isolation. AI can help strengthen those bonds by making it easier to share memories, stay present, and create new experiences together, even across distance.

How AI Brings Families Closer

Photo Sharing: AI-powered photo apps organize pictures by faces, making it easy to create family albums or slideshows.

Shared Projects: Grandparents and grandchildren can write stories together, with AI filling in fun details.

Voice Memories: AI can record and transcribe family conversations, preserving stories for future generations.

Video Calls Made Simple: AI smooths connections and keeps everyone in the frame so moments feel more natural.

Everyday Examples

A granddaughter sits with her grandmother and uses AI to ask, "Write a fairy tale about a brave cat named Whiskers." They giggle as the story unfolds.

A grandfather shares a restored black-and-white photo of his wedding day with the whole family group chat. AI brought the colors back to life.

A caregiver helps set up a family "shared calendar," so everyone knows about birthdays, doctor visits, or Sunday dinners.

Caregiver's Corner

For those guiding these connections:

Set Up Shared Spaces: Create family group chats, shared photo albums, or digital calendars.

Teach One Feature at a Time: Too many tools at once can overwhelm. Start with photo albums or video calls.

Encourage Participation: Invite seniors to be the "storytellers" in AI family projects, letting them guide while younger members handle the tech.

Celebrate Togetherness: Every successful shared activity is worth celebrating. The point is bonding, not perfection.

Try This: Family Story Game

1. Open an AI chat app.
2. Type or say: "Write a story about [grandchild's name] and Grandma going on an adventure."
3. Read the story aloud together.
4. Add your own twists and endings as a family.
   Not only does this spark laughter, but it turns technology into a shared playground across generations.

Reassurance: Technology Can Be a Bridge, not a Barrier

Instead of creating distance, AI can help families find new ways to connect. Whether through stories, photos, or simple conversations, technology becomes the glue that holds generations closer, strengthening the bond between seniors, caregivers, and loved ones.

In this chapter, you saw how AI can help bring families together in meaningful ways. In the next chapter, we'll focus on something deeply personal: how seniors can use AI to leave a legacy and preserve their wisdom for the future.

NOTES

8 CHAPTER: LEAVING A LEGACY

Every life is a story worth telling. Seniors carry decades of wisdom, experiences, and family history that younger generations may never fully know unless it's preserved. AI can be a powerful ally in capturing those treasures. From memoir writing to organizing photo albums, technology makes it easier than ever to leave a legacy.

How AI Helps Preserve Memories

Memoir Writing: AI tools can prompt seniors with questions ("Tell me about your first job" or "What was your wedding day like?") and help shape their answers into stories.

Photo Restoration: Old, faded photos can be restored, colored, and organized with the help of AI.

Voice Preservation: AI can record and transcribe conversations, turning spoken words into written keepsakes.

Family History Projects: AI can assist with genealogy research, tracing ancestors, and connecting family trees.

Everyday Examples

A grandmother records her childhood memories into a phone app, which transcribes them into a booklet her family can keep forever.

A senior uploads old family photos into a digital album, where AI automatically groups them by faces and years.

A caregiver helps a grandfather use AI to generate a family recipe book complete with stories about each dish.

Caregiver's Corner

For those helping seniors preserve their legacy:

Prompt Gently: Ask simple questions that spark storytelling. "What was your favorite childhood game?"

Use AI as a Helper, Not an Editor: Let seniors' authentic voices shine. AI can clean up grammar, but personality is what matters most.

Make It a Ritual: Set aside regular times for "story sessions" with AI tools running in the background.

Share the Results: Print stories, make albums, or create recordings to gift family members.

Try This: Memoir Prompt

1. Open a notes or AI writing app.
2. Type or say: "Tell me about a time you felt the happiest."
3. Answer the question in your own words.
4. Watch as the AI organizes your story into clear sentences.

   Do this regularly, and soon you'll have a collection of stories that tell your life's journey.

Reassurance: Your Voice Matters

AI doesn't replace your story; it helps make sure your story lives on. Every memory you share, every photo you save, every word you speak becomes part of the legacy your family can hold close for generations to come.

In this chapter, you learned how AI can preserve memories and wisdom as a living legacy. In the next chapter, we'll focus on building confidence in learning and using AI without fear.

NOTES

9 CHAPTER: LEARNING WITHOUT FEAR

For many seniors, the hardest part of using technology isn't the buttons or the apps, it's the fear. Fear of "messing up," fear of "looking silly," or fear of being left behind. But here's the truth: learning AI doesn't require perfection. It requires only curiosity, patience, and a willingness to try.

The Mindset Shift

From Fear to Curiosity: Instead of worrying, "What if I break it?" ask, "What might I discover?"

From Pressure to Play: Treat learning AI like experimenting, not like passing a test.

From Alone to Supported: Remember, caregivers, family, and friends are here to guide you.

Why Mistakes Are Part of the Journey

Pressing the wrong button won't ruin your phone. Deleting a message doesn't erase your memories. Technology has built-in safety nets, and if something goes wrong, most problems can be fixed with a tap, restart, or call for help. Every mistake is simply a step toward learning, not a sign of failure.

Caregiver's Corner

For those guiding seniors:

Celebrate Progress: Even small wins like sending a text or setting a reminder are victories.

Avoid Overloading: Introduce one feature at a time. Too much at once fuels frustration.

Reframe Mistakes: Say, "That's how we learn," instead of "That's wrong."

Offer Reassurance: Remind seniors they're not falling behind, they're adapting at their own pace.

Try This: One New Skill

1. Pick one small feature you've never used before (like asking a voice assistant a question).
2. Try it once, don't worry about mistakes.
3. Repeat it the next day.
4. Notice how much easier it feels already. Confidence grows step by step.

Reassurance: You Are Never Too Old to Learn: Age doesn't close the door to learning; it opens new ways to apply wisdom. You bring patience, resilience, and life experience to the table. AI is just a new tool, and like any tool, you can master it with time and support.

In this chapter, you've seen how to move past fear and build confidence at your own pace. In the final chapter, we'll look ahead with hope, exploring how AI may continue to grow and how seniors and caregivers can thrive together in a friendly future.

NOTES

10 CHAPTER: THE FUTURE IS FRIENDLY

When people talk about the future of technology, it can sound overwhelming: robots, self-driving cars, smart homes that do everything for you. But here's the most important thing to remember: the future is not something to fear, it's something you can grow into, one step at a time.

AI isn't here to take away your independence or your humanity. It's here to support you, to make daily life easier, and to help you stay connected with the people you love.

What's Coming Next (In Simple Terms)

Health Innovations: AI will help doctors catch illnesses earlier, suggest better treatments, and even monitor health from home.

Smarter Homes: From lights that turn on when you enter the room to appliances that remind you when food is low, AI can make homes safer and more comfortable.

Transportation: Self-driving cars and AI-assisted vehicles will open doors for seniors who may no longer drive but still want mobility.

Companion Tools: Friendly AI companions,

such as small devices or apps, may provide comfort, reminders, and even conversation.

Everyday Hope

Imagine asking your watch to check your heart rate, your speaker to call your grandson, and your tablet to show you old family photos all in a matter of seconds. That future is already here, and it will only grow easier, faster, and friendlier.

Caregiver's Corner

For those guiding seniors into the future:

Model Optimism: Show excitement, not dread, when new features arrive.

Keep Learning Together: Seniors and caregivers can explore updates as a team.

Focus on Benefits, Not Fears: Highlight what

technology adds to life: independence, safety, joy.

Remember Humanity: AI is a tool, but empathy, love, and patience are still the greatest supports you can offer.

Try This: A Glimpse into Tomorrow

1. Ask a voice assistant: "What's something new AI can do today?"
2. Listen to the answer.
3. Reflect together: How could this make life easier or more fun?
   Small glimpses into what's ahead can spark excitement instead of fear.

Reassurance: The Best Is Yet to Come

You are not too old to learn, and you are not

too late to benefit from this wave of technology. The future is not racing away without you; it's waiting to walk alongside you. AI is not about replacing the human touch; it's about amplifying it, supporting it, and making sure you can thrive with confidence and joy.

With this final chapter, we close the handbook with the same message we began: AI is not here to scare you. It is here to help you. Together, seniors and caregivers can explore this new digital age with curiosity, courage, and hope.

NOTES

11 CHAPTER: HOW TO FIND GEN AI TO USE

A Friendly Start

By now, you know that artificial intelligence can be a helpful companion, from reminding you to drink water to connecting you with family to adding joy to your daily life. But many seniors and caregivers still wonder: "Where do I even find these AI helpers?"

This chapter gives you simple, step-by-step directions to start using generative AI apps safely and with confidence.

What Are Generative AI Apps?

Generative AI apps are special programs that can create new things when you ask for stories, poems, pictures, songs, or even answers to everyday questions. They listen to your instructions, called prompts, and generate something new in response.

Some popular examples are:

ChatGPT – answers questions, explains things, and helps with writing.

Google Gemini – helpful for quick

information and explanations.

Microsoft Copilot – works with Microsoft tools and the Edge browser.

Claude (Anthropic) – designed for clear, thoughtful conversations.

Perplexity AI – acts like a search engine that explains its answers.

Where to Find Them

On the Web (Computer or Tablet Browser)

Open your internet browser (Chrome, Safari, or Edge) and type these addresses directly into the top bar:

chat.openai.com → ChatGPT

gemini.google.com → Google Gemini

copilot.microsoft.com → Microsoft Copilot

claude.ai → Claude

perplexity.ai → Perplexity

Tip: Bookmark the page so you can return with just one click.

On a Phone or Tablet (Apps)

iPhone / iPad: Open the App Store, search for "ChatGPT" or "Gemini," then tap Download.

Android devices: Open the Google Play Store, search for the app, and tap Install.

Always choose the official app (look for the company name: OpenAI, Google, Microsoft, Anthropic).

Getting Started Safely

Step 1: Go to the official website or download from the official store.

Step 2: Create an account using your email and a strong password.

Step 3: Write down your login details and keep them somewhere safe.

Step 4: Bookmark the site or pin the app to

your home screen.

Step 5: Try a first simple prompt, such as:

o "Tell me a cheerful short story about a cat."

o "Give me three tips for staying active indoors."

Free vs. Paid Versions

Most apps have a free version, which is plenty to get started.

Paid versions may respond faster or give access to the newest tools, but there's no rush.

Caregiver Tip: Begin with free exploration first before deciding if upgrades are useful.

Caregiver's Corner

Sit with your loved one during setup, help with typing or downloading if needed.

Encourage them to try fun prompts first, like

"Write a short poem about my favorite flower."

Print a simple "cheat sheet" with login steps and common prompts.

Show how to return to the app easily (bookmark or shortcut).

Try This Together

Open ChatGPT or Gemini and type:

"Tell me a joke about grandparents and grandkids."

"What's a simple soup recipe I can make with carrots and potatoes?"

"Explain how to change a TV remote battery in very simple steps."

Reassurance

Remember: You cannot "break" these apps by asking the wrong question. They are built to respond politely, even if you make a mistake or typo. Think of it like learning to ride a bike, wobbly at first, but each try builds confidence. Your first steps into generative AI don't need to be perfect; they just need to begin.

NOTES

GLOSSARY

A

Accessibility Features – Special tools built into devices that make them easier to use, such as larger text, magnifiers, voice commands, or screen readers that read words out loud.

Algorithm – A set of step-by-step instructions that a computer follows to solve a problem or make a decision. For example, when YouTube suggests a video, it's following an algorithm.

App (Application) – A program you can install on your phone, tablet, or computer. Examples: Facebook, Zoom, medication reminder apps.

Artificial Intelligence (AI) – Technology that allows computers to "learn" from information and perform tasks like recognizing speech, suggesting movies, or answering questions.

Augmented Reality (AR) – A technology that

adds digital images or information on top of the real world, usually through a phone camera. Example: an app showing how furniture would look in your living room.

B

Biometrics – Using unique features of your body (fingerprint, face scan, or voice) to unlock a device or verify identity.

Blockchain – A digital "record book" that is secure and very difficult to change. Most often used for cryptocurrency.

Bot – A computer program that runs automatically. Some are helpful (like customer service chatbots); others spread spam or scams.

Browser – The program you use to access the internet, such as Chrome, Safari, or Edge.

C

Captcha – A short test (like typing letters from a picture or clicking on images) that proves you are human, not a bot.

Chatbot – A program powered by AI that talks with you through text or voice. Example: the little chat window that pops up on a website.

Cloud (The Cloud) – Storing files, photos, or programs on the internet instead of just on your personal device. Example: Google Photos or iCloud.

Cookie (Internet Cookie) – A small file saved by websites to "remember" you, like keeping you logged in or showing you ads related to past searches.

Cybersecurity – The practice of keeping devices, accounts, and information safe from hackers or scammers.

D

Data – Information collected by computers, such as your photos, search history, or steps tracked on a fitness device.

Deepfake – A video, photo, or audio clip made with AI to look real but is actually fake. Example: a video of someone saying words they never said.

Device – Any piece of electronic equipment, such as a phone, tablet, or computer.

Digital Divide – The gap between people who use technology easily and those who don't have access or skills.

Digital Footprint – All the traces you leave online: websites you visit, posts you share, or photos uploaded.

Digital Wallet – An app that stores payment info so you can pay with your phone instead of using cash or cards.

E

Email Phishing – Fake emails designed to trick you into giving away personal info.

Encryption – A way of locking up digital information so only the right people can unlock it. Used by banks and health apps.

Emoji – Small digital pictures used in texts or online, such as 😀 ❤️ 👍.

E-Reader – A device (like Kindle or Nook) designed specifically for reading digital books.

F

Facial Recognition – Technology that identifies people by scanning their faces. Often used to unlock smartphones.

Fake News – Articles, posts, or videos that look real but are designed to mislead.

Firewall – A protective "wall" in your computer that blocks harmful internet traffic.

Fitness Tracker – A wearable device (like Fitbit or Apple Watch) that counts steps, tracks heart rate, or monitors sleep.

G

Generative AI – AI that creates new things such as text, pictures, or music. Examples: ChatGPT (text), DALLE (images).

GPS (Global Positioning System) – Satellite technology that helps phones and cars know their exact location. Used for maps and directions.

H

Hack / Hacker – A person who breaks into a computer or account without permission, often to steal info or cause harm.

Hashtag (#) – A word or phrase with the "#" symbol used to group posts online. Example: #FamilyPhotos or #Travel.

Hotspot – A way to connect devices to the internet using Wi-Fi from your phone or another device.

I

Identity Theft – When someone steals personal information (like Social Security number) to commit fraud.

Influencer – Someone online who has a large following and can affect opinions or trends.

Internet of Things (IoT) – Everyday objects (like thermostats, fridges, or doorbells) that connect to the internet.

M

Malware – Harmful software designed to

damage a computer or steal information. Includes viruses and spyware.

Memory Aid Apps – Programs that help with reminders, notes, or daily routines.

Misinformation – False or misleading information, often spread by accident.

Machine Learning – A type of AI that allows computers to "learn" and improve from data. Example: spam filters learning to catch new junk mail.

N

Neural Network – A design in AI inspired by the human brain, used to recognize patterns like voices, handwriting, or photos.

Notification – A message that pops up on your device, such as an email alert, calendar reminder, or text message.

P

Password Manager – A secure program that stores all your passwords and helps create strong ones.

Personal Data – Information about you such as your name, birthday, address, phone number, or medical info.

Phishing – A scam where someone pretends to be trustworthy to steal personal info.

Podcast – An audio program you can listen to on demand, like a radio show.

Privacy Settings – Controls that let you decide what info websites, apps, or devices can collect and share.

QR Code – A square barcode you can scan with your phone camera to open a website, menu, or app.

R

Recommendation Engine – The system that suggests things online, like "You may also like…" on Amazon or Netflix.

Robocall – An automated phone call, often a scam or advertisement.

Robot – A machine that can perform tasks automatically, sometimes with AI. Examples: Roomba vacuum or hospital delivery robots.

S

Scam – A trick meant to steal money or information, often through fake calls, texts, or websites.

Screenshot – A picture taken of whatever is showing on your screen.

Smart Device – Any device that connects to the internet and can "think" a little. Examples:

smart TVs, smart thermostats, or smartwatches.

Smart Home – A house with internet-connected devices, like lights, locks, and appliances, that can be controlled remotely.

Social media – Websites and apps where people share and connect, like Facebook, Instagram, or TikTok.

Spam – Unwanted or junk emails, texts, or calls.

Streaming – Watching or listening to media directly from the internet without downloading it. Examples: Netflix, YouTube, Spotify.

T

Tablet – A handheld device bigger than a phone but smaller than a laptop, like an iPad.

Two-Factor Authentication (2FA) – An extra step for security: after entering your password, you also enter a code sent to your phone.

Troll – Someone who posts rude or false comments online to upset others.

V

Virtual Assistant (Voice Assistant) – An AI program that responds to spoken commands, like Siri, Alexa, or Google Assistant.

Virtual Reality (VR) – A computer-made world you can experience with special goggles.

Virus (Computer Virus) – A harmful program that spreads between devices and can cause damage.

Voice-to-Text – Technology that types what you say out loud, useful if typing is difficult.

W

Wearables – Devices worn on the body, like smartwatches, fitness trackers, or Bluetooth-enabled hearing aids.

Wi-Fi – A wireless internet connection that lets your devices go online without plugging in.

Wireless Charging – A way of charging devices by placing them on a special pad instead of plugging them in.

NOTES

QUICK SAFETY REFERENCE

Top 10 Safety Tips

1. **Pause Before Clicking** – If something feels odd, don't open it.
2. **Never Share Personal Info** – Banks and Social Security will *never* ask for passwords or Social Security numbers by email/text.
3. **Check the Source** – Look at the sender's email or phone number carefully.
4. **Strong Passwords** – Use phrases like *SunnyGarden1955!* instead of 1234.
5. **Use Two Steps (2FA)** – Extra code sent to your phone adds protection.
6. **Beware of Urgency** – Scams often say, "Act now!" or "Emergency!" to trick you.
7. **Don't Trust Unknown Links** – If unsure, don't click it. Ask a trusted person first.

8. **Update Devices** – Keep phones and computers updated for the latest safety.
9. **Use Trusted Apps Only** – Download from official app stores (Apple, Google).
10. **Ask for Help** – Caregivers, family, or friends can check before you act.

Key Words to Know

- **Phishing:** Fake messages pretending to be from someone you trust.
- **Spam:** Junk emails, texts, or calls you don't want.
- **Scam:** A trick to steal money or information.
- **Password Manager:** A tool that safely remembers all your passwords.
- **Wi-Fi:** Wireless internet connection at home, coffee shops, or libraries.

Keep this card near your computer, phone, or tablet.

If something feels confusing, **STOP, THINK, ASK** before acting.

NOTES

LEARN MORE

This section offers trusted resources to help seniors and caregivers continue learning, stay safe, and explore AI tools at their own pace.

Getting Started with Technology

- **Senior Planet (AARP):** Free online classes and workshops for older adults. www.seniorplanet.org
- **AARP Technology Help:** Step-by-step guides on smartphones, apps, and staying safe online. https://www.aarp.org/home-family/personal-technology
- **Oasis Connections:** Technology training designed for adults 50+. https://www.oasisnet.org/connections

---

AI Basics

- **What is Artificial Intelligence? (MIT):** A plain-language guide. https://mitsloan.mit.edu/ideas-made-to-matter/what-is-artificial-intelligence
- **Common Uses of AI in Everyday Life (IBM):** Examples of how AI already helps. https://www.ibm.com/topics/artificial-intelligence
- **AI for Beginners (Google AI):** Introductory tools and FAQs. https://ai.google/education

---

Safety & Security

- **FTC: Online Safety for Older Adults:** Guidance to avoid scams and fraud. https://www.consumer.ftc.gov
- **Cyber-Seniors:** Training and mentorship

for seniors on internet safety. www.cyberseniors.org

- **StaySafeOnline (National Cybersecurity Alliance):** Simple checklists for privacy and security. https://www.staysafeonline.org

---

Health & Wellness with AI

- **Mayo Clinic: AI in Medicine:** How AI supports patient care. https://www.mayoclinic.org/ai
- **National Institute on Aging:** Resources for seniors' health and caregiving. https://www.nia.nih.gov/health
- **MedlinePlus:** Trusted information on health apps and tools. medlineplus.gov

---

Fun & Creativity

- **Smithsonian Open Access:** Explore millions of free images, artifacts, and history. https://www.si.edu/openaccess
- **StoryCorps:** Record and share personal stories with family or future generations. https://www.storycorps.org
- **AI Playgrounds:** Try safe AI creative tools like image generators or storytelling apps recommended by caregivers.

Books & Further Reading

- Coughlin, J. F. (2017). The Longevity Economy: Unlocking the World's Fastest-Growing, Most Misunderstood Market. PublicAffairs.
- Gelles, D. (2022). *The Man Who Broke Capitalism*. Simon & Schuster. (Includes discussion on technology shifts).

- Tegmark, M. (2017). Life 3.0: Being Human in the Age of Artificial Intelligence. Vintage.

- **Tip:** If online links feel overwhelming, ask a caregiver, librarian, or trusted friend to help you explore them. Libraries often run free "tech help" sessions too.

ABOUT THE AUTHOR

Alice Cuthbertson is an author, speaker, technology leader, and founder of Aliboo Publishing. With a career spanning technology, program management, leadership, and artificial intelligence, she is passionate about helping people understand and embrace technology in ways that are practical, ethical, and human-centered.

As the daughter of a nurse, Alice spent much of her childhood visiting nursing homes and interacting with seniors from all walks of life. Those early experiences instilled a deep appreciation for the wisdom, resilience, and stories of older generations. They also inspired her belief that technology should be accessible to everyone, regardless of age or technical experience.

Alice holds degrees in Biological Sciences and Information Technology and is pursuing doctoral research focused on leadership, empathy, and the responsible use of artificial intelligence. She is also a certified project management professional, Agile practitioner, AI ethics advocate, and IBM Blue Core Coach.

She is the author of *The AI Parenting Handbook: Raising Kids in the Digital Age* and *The AI Handbook: For Seniors and Their Caregivers*, both part of the AI Parenting Handbook Series.

Speaking Engagements

Alice regularly presents on topics including:

Artificial Intelligence for Seniors and Caregivers

Parenting in the Age of AI

Responsible and Ethical AI Use

Human-Centered Technology

Leadership and Empathy in an AI-Driven World

Practical AI for Everyday Life

Her presentations are designed for senior communities, retirement organizations, libraries, churches, community groups, schools, caregiver networks, and professional associations. Sessions are available both in person and virtually and can be customized for audiences of all sizes.

For speaking engagements, workshops, book clubs, interviews, or community events, please contact:

Aliboo Publishing

Email: Alice @ aliboo@aliboo.com

*"*Technology is most powerful when it helps us stay connected, preserve what matters, and bring out the best in one another.*"*

www.ingramcontent.com/pod-product-compliance
Lightning Source LLC
LaVergne TN
LVHW022008200926
848481LV00007B/2292
* 9 7 9 8 9 9 5 9 0 3 2 1 5 *